Glencoe

Outline Map

Resource Book

 Glencoe McGraw-Hill

New York, New York   Columbus, Ohio   Woodland Hills, California   Peoria, Illinois

Glencoe/McGraw-Hill

A Division of The **McGraw·Hill** *Companies*

Send all inquiries to:
Glencoe/McGraw-Hill
8787 Orion Place
Columbus, Ohio 43240-4027

ISBN 0-07-824996-1

Printed in the United States of America

4 5 6 7 8 9 10 047 08 07 06 05 04 03

Contents

Introduction

Outline maps are among the most valuable tools available to social studies teachers today. Their use is limited only by the imaginations of individual teachers. All areas of the social studies curriculum can benefit by using the outline maps.

The outline maps provide a visual image of information. Studies have shown that most students learn better when they can visualize material. The outline maps provide a context for understanding the world in spatial terms. As students add details to the basic outline maps, they are actively organizing information in spatial terms. They can explore the physical characteristics of places such as land-forms, bodies of water, and vegetation. Students enjoy using outline maps as hands-on activities that stimulate interest and provide variety in the classroom. Outline maps invite students to create their own learning materials. Students can use the maps they create as they prepare for various assessments.

The outline maps provided in this book are valuable learning aids because of their flexibility. They are inexpensive and easy to use. The same map can be used for several topics in one chapter, or even for several different courses. Used creatively by the teacher, outline maps can be invaluable classroom tools.

Uses for Outline Maps

Outline maps are teaching aids that can broaden classroom experiences. How can teachers best determine ways to use these tools? The following information suggests a few ways to approach the use of outline maps in social studies classrooms.

Practicing Map Skills

Recent reports note an alarming decline in the ability of students to use basic map skills. These basic skills include using scales and legends, determining longitude and latitude, reading relative and exact location, and identifying symbols. Standardized tests are now requiring that students acquire some degree of map knowledge.

Rather than placing the entire burden of teaching map skills on geography teachers, all social studies teachers can help in this effort. All social studies teaching can be enhanced through the use of outline maps.

Developing Perspectives

As visual tools, maps help students develop perspective. Using maps, students can learn basic personal perspectives. They can identify where their city, county, state, and country are located within the broader world picture.

Then, relationships and higher-level concepts can surface. Using maps to teach a history lesson on the Cold War, students can see how far their homes are from the former Soviet Union. They may also realize how close Alaska is to the former Soviet Union. They can then begin to understand concepts such as "spheres of influence."

Students can use outline maps to compare and contrast land areas and bodies of water. They can also analyze location and distances. They can explore such questions as: How does the size of the United States compare with the size of Australia? How close is the southern tip of Chile to the South Pole?

Noting Patterns and Trends

Maps can put complex statistics and factual information into a readily understandable form. Statistics about the amounts and locations of oil and natural gas reserves throughout the world might leave students without a clear picture of what those figures mean. A resource map would clearly show students that (a) the countries of the Middle East have the largest reserves, and (b) there are some places in the world that are completely lacking in those resources.

A list showing Electoral College votes by state for each presidential candidate gives the basic facts. A map that uses color to show which states each candidate won expresses the election results in an effective way. Using such maps for several elections, students can readily see voting patterns emerge.

Recognizing Change

Another difficult concept for students is the idea of change. Politics, the environment and society are constantly shifting. Outline maps are perfect for showing "before" and "after" pictures. A teacher might use an outline map to show changing political borders, such as those of Poland before and after World War II. An outline map could be used to show the shifting of the American population to the Sunbelt areas of our country. Maps can also be used to show geographical changes, such as the erosion of seashores by the oceans or the alteration of landscapes by glacier movements.

Perfecting Social Studies Skills

Besides helping students learn specific course content, maps can help students perfect their social studies skills. For example, once information about the distribution of natural resources throughout the world is pictured on a map, students can be asked to draw conclusions of their own.

Some students may suggest that resources are unevenly distributed and, therefore, must be shared among countries. Others may realize that many resources are available in areas of the world that are undeveloped, therefore suggesting the need for exploration. Drawing conclusions, a critical thinking skill, is only one example of the skills developed by discovery learning. Other skills that can be developed using outline maps include making inferences, making generalizations, classifying information, predicting consequences, and evaluating information.

Selected Activities for Outline Maps

World History

1. Locate and label the centers of early civilizations in Mesopotamia, Egypt, India, China, Africa, and the Americas.
2. Create thematic maps representing various aspects of world history such as the spread of Greek and Roman civilizations; major trade routes that linked Europe, the Islamic world, and East Asia; and the rise and fall of the great African empires such as Ghana, Mali, and Songhai.
3. Locate and label places and regions of historical significance such as the Indus, Nile, Tigris and Euphrates, and Yellow (Huang He) River valleys, and describe their physical characteristics.
4. Locate and analyze the effects of human geography, such as the impact the building of the Suez and Panama Canals had on world trade patterns.
5. Use maps to interpret and explain geographic factors such as how the control of the Strait of Hormuz has influenced people and events in the past.
6. Explore the changes that have occurred in the countries of eastern Europe during the twentieth century.

American History

1. Trace principal voyages of exploration from Europe to the Americas by such adventurers as Columbus, Vespucci, Cabot, Hudson, and Cartier.
2. Create thematic maps representing various aspects of American history such as identifying the areas of the Americas claimed and settled by the world powers active in the exploration and colonization of the Western Hemisphere and the changing transportation patterns that resulted from the interstate highway system.
3. Create maps showing changes in political boundaries that resulted from statehood and international conflicts such as the territories granted to the United States at the end of various wars.
4. Identify changes in population and the physical features of the country that influenced these changes.
5. Trace major military campaigns during wars such as the French and Indian War, the Revolutionary War, and the Civil War.
6. Outline and label the major territorial acquisitions of the United States from 1776 to the present.
7. Identify the order in which states were granted statehood.
8. Trace the routes followed by adventurers such as Lewis and Clark.
9. Identify locations of military conflicts such as Vietnam, Korea, the Persian Gulf, and Bosnia and Herzegovina.

Government and Civics

1. Locate physical features that are affected by government policies in various regions or places.
2. Use color or shading to compare the seats in the House of Representatives before and after the 2000 census.
3. Use color and labeling to show the percentage of voting age population by state that participated in presidential elections.
4. Use color and labeling to show the state-by-state results of presidential elections. Identify any notable voting patterns by region.
5. Identify changes in population among the states and the effect of these changes on government policies.
6. Identify states voting "for" or "against" a constitutional amendment. Identify any notable regional voting patterns.
7. Identify and label the percentages of ethnic or cultural groups found in various states or regions.

Geography

1. Locate and label major ocean currents and wind patterns on a world map.
2. Use distinctive colors to show population densities in various regions of the world.
3. Outline and label the seven physical regions of the United States.
4. Locate and label major landforms such as mountain ranges, deserts, and bodies of water.
5. Locate and label countries and their capitals.
6. Label various physical regions of the world by climate and vegetation.
7. Label various regions of the world by human factors such as language and religion.
8. Locate and label major historical and contemporary societies.
9. Create thematic maps representing such things as population distribution, economic activities, and natural resources.

Economics

1. Outline and number the regions of the United States Federal Reserve districts. Then, locate the bank and branch cities within each region.
2. Label the general sales tax rate for each state of the United States.
3. Use color to identify industrial and non-industrial countries of the world.
4. Locate and label the natural resources available in various regions of the world, and analyze the impact of natural resource allocation on the wealth of countries in the region.
5. Locate and label the literacy rates of various countries in a region.
6. On a world map, use color to identify developed and developing countries of the world.
7. Use colors to identify the levels of imports to and exports from the United States.

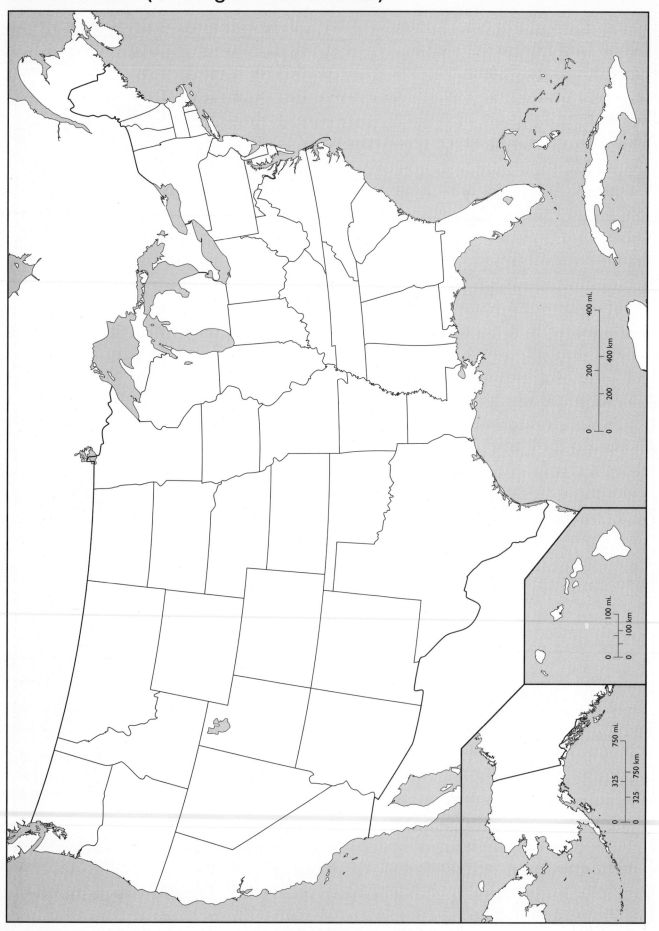

UNITED STATES (including Alaska and Hawaii)

400 mi.
200
200
0
400 km
400
200
0

100 mi.
0
100 km
0

750 mi.
325
325
0
750 km
750
0

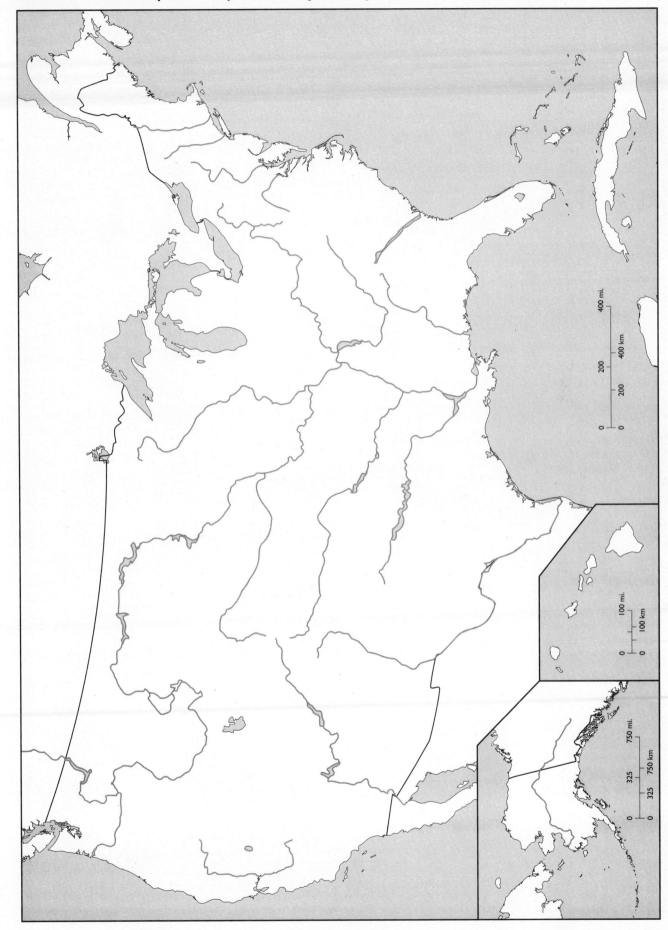

UNITED STATES (with major river systems)

UNITED STATES (with Alaska and Hawaii correctly placed)

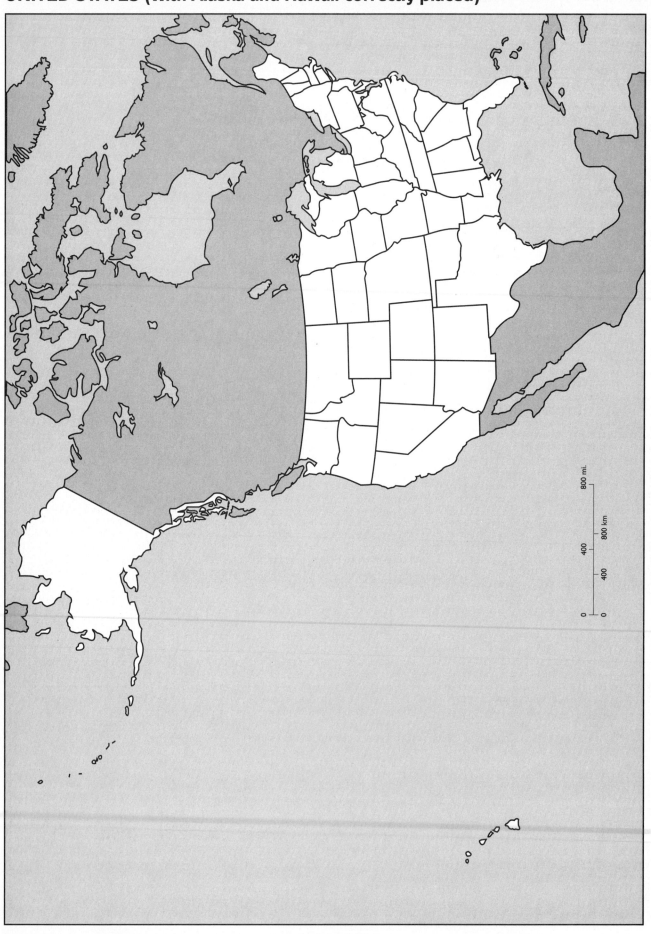

800 mi.

800 km

400

400

0   0

THIRTEEN COLONIES

0   200   400 mi.

0   200   400 km

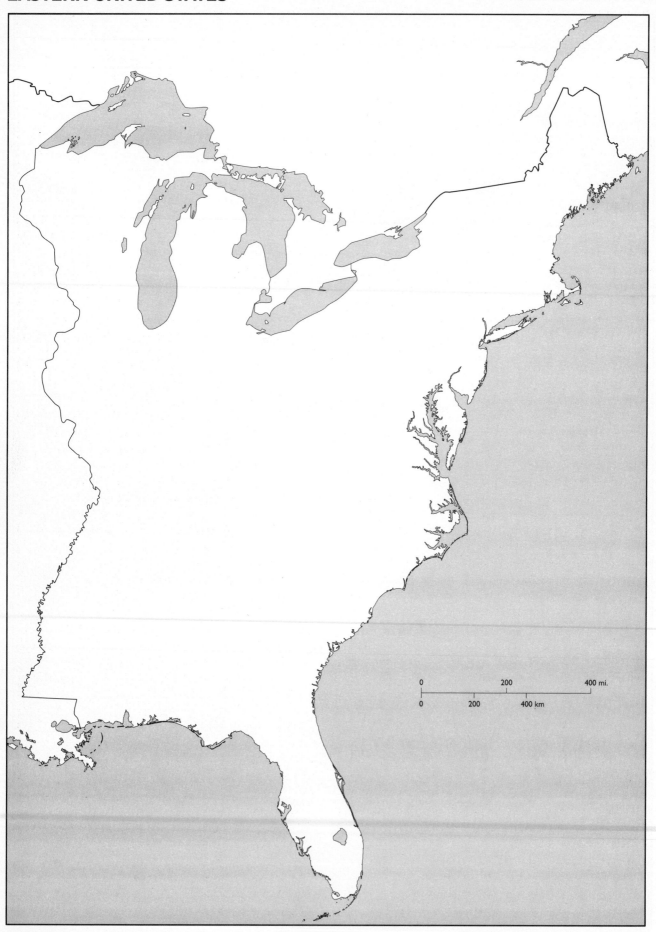

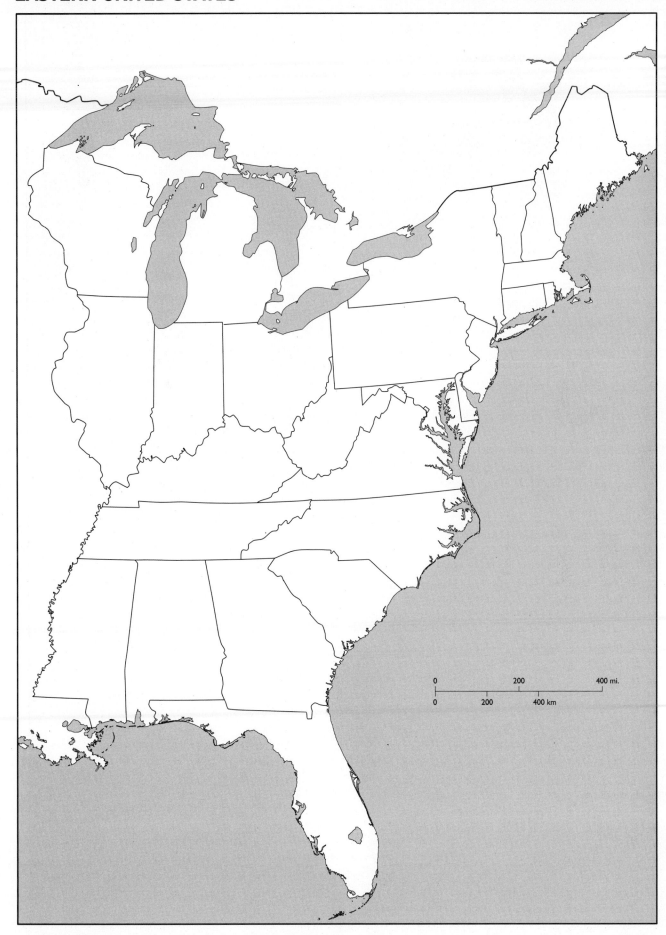

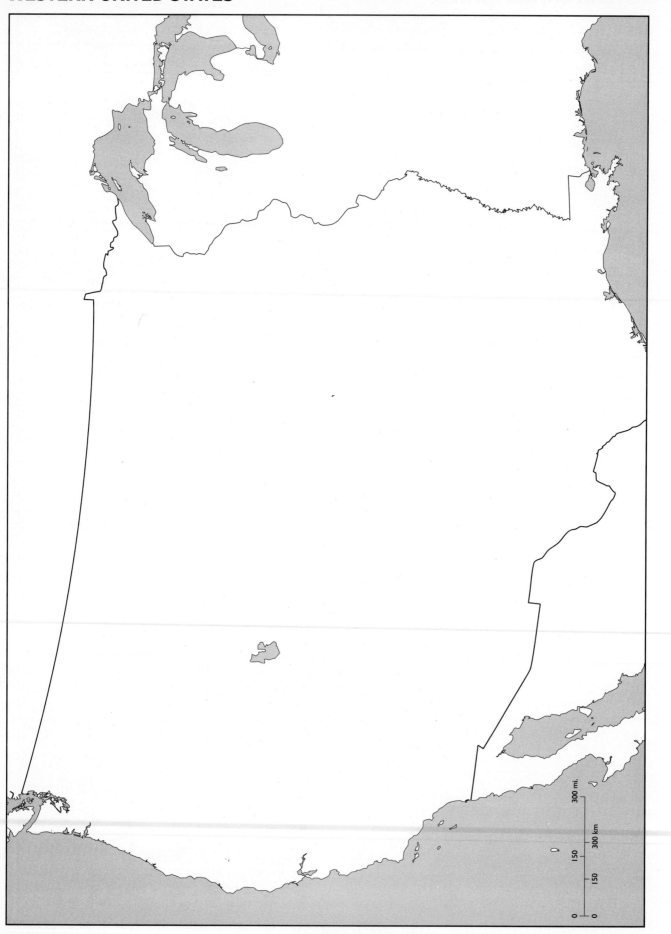

300 mi.

300 km

150

150

0   0

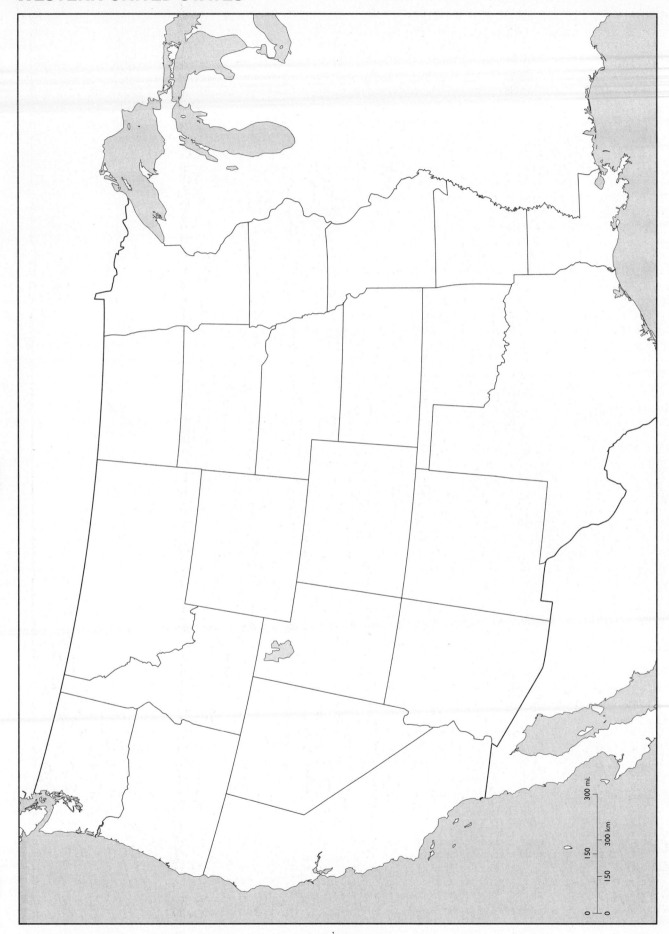

300 mi.

300 km

150

150

0

0

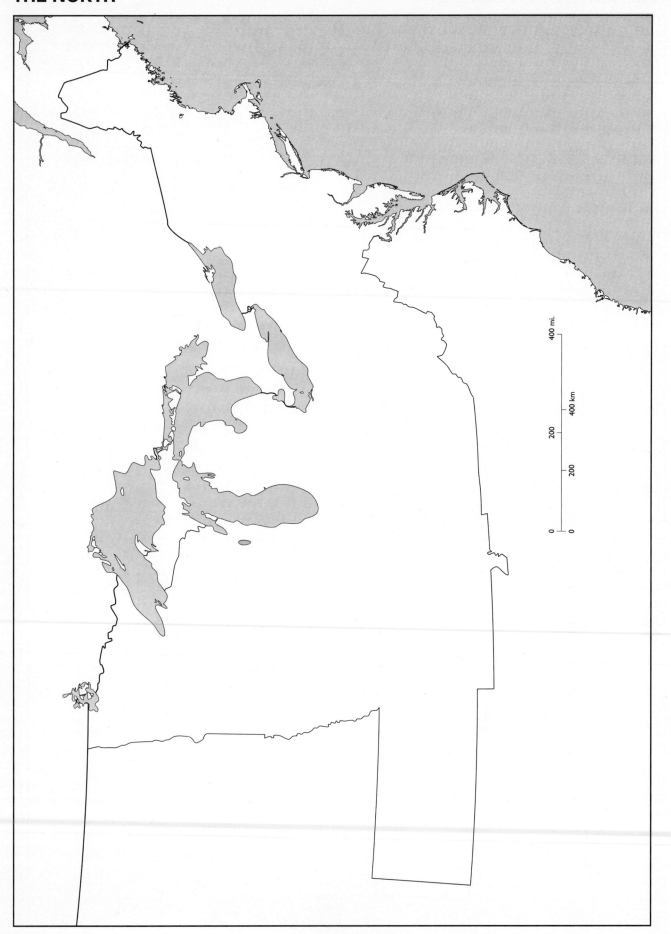

400 mi.

400 km

200

200

0

0

THE NORTH

400 mi.

400 km

200

200

0

0

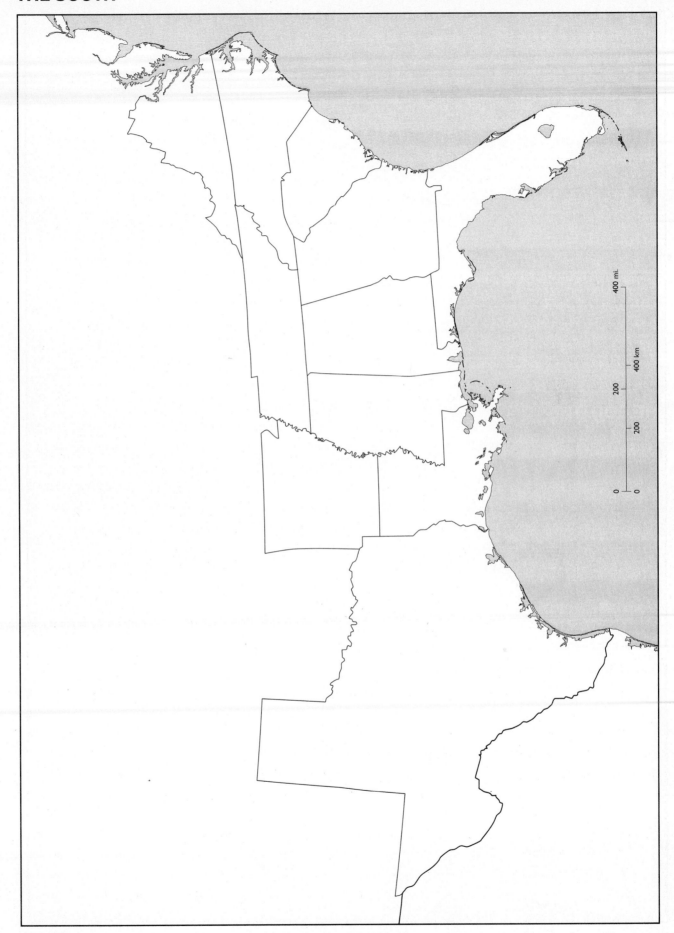

THE SOUTH

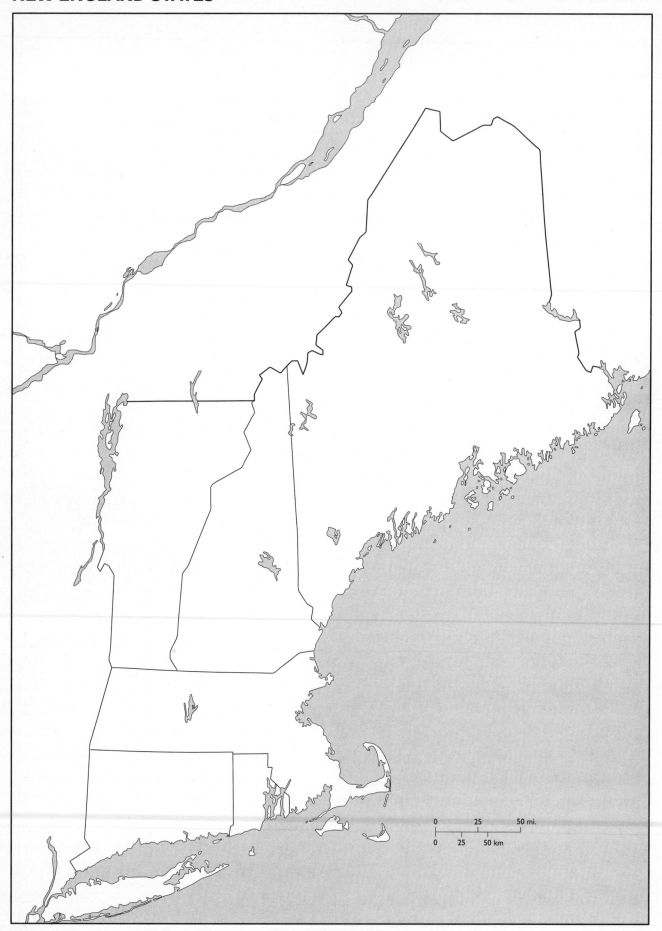

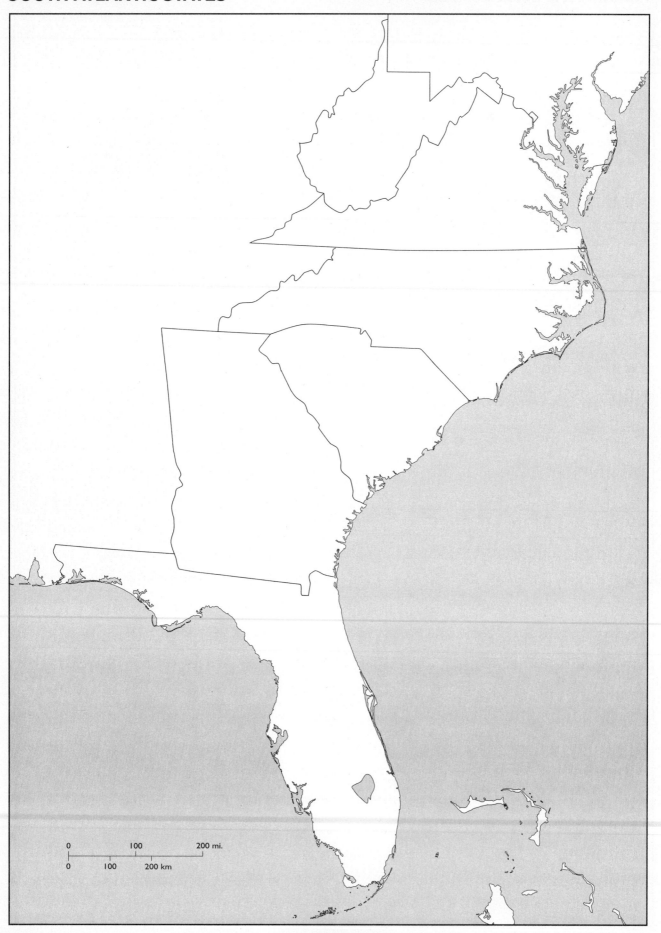

0   100   200 mi.

0   100   200 km

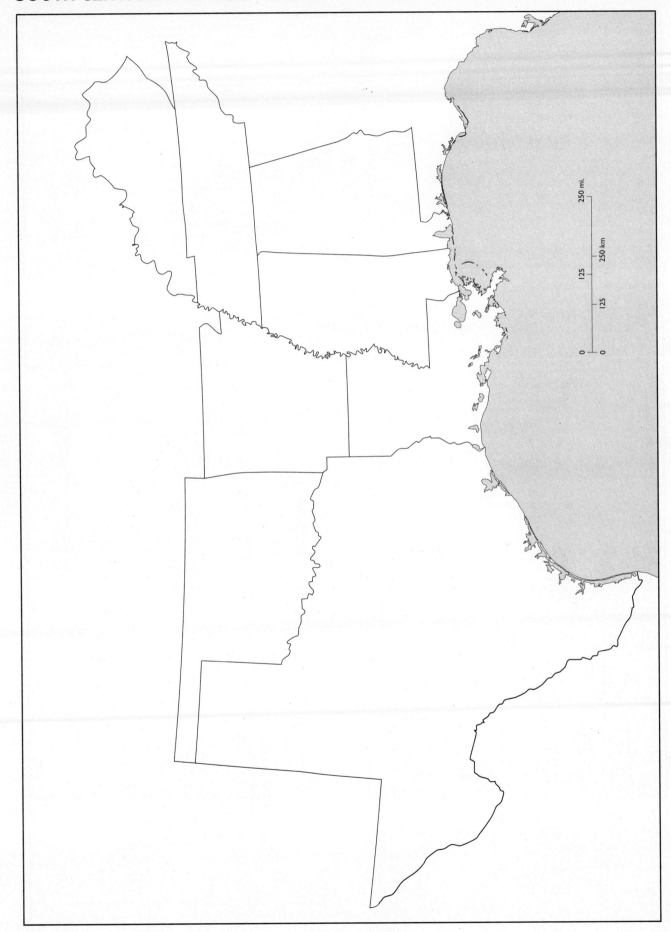

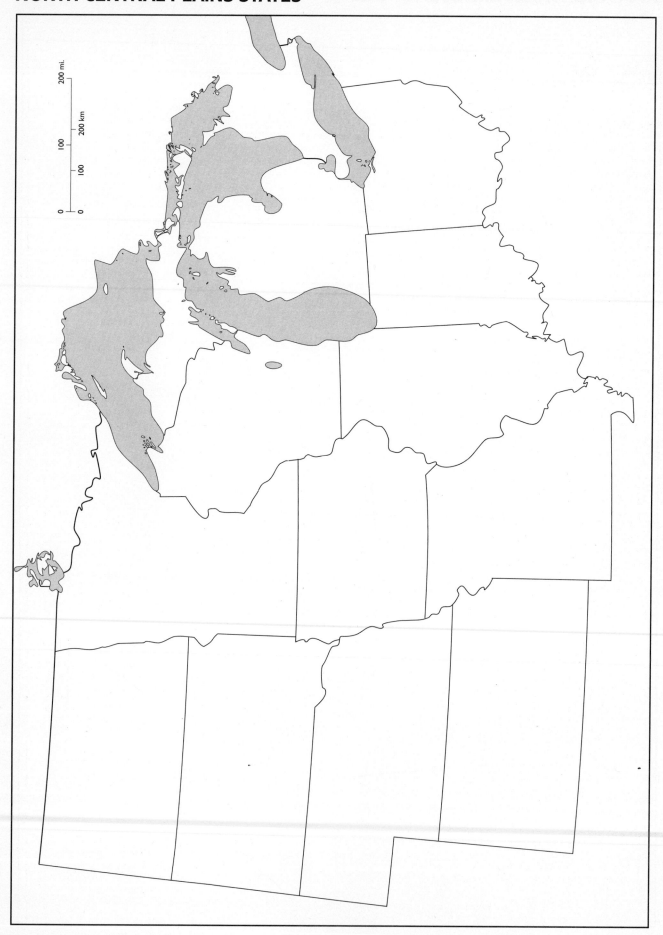

200 mi.

200 km

100

100

100

100

0

0

ROCKY MOUNTAIN AND PACIFIC STATES

0 200 400 mi.

0 200 400 km

NORTH AMERICA

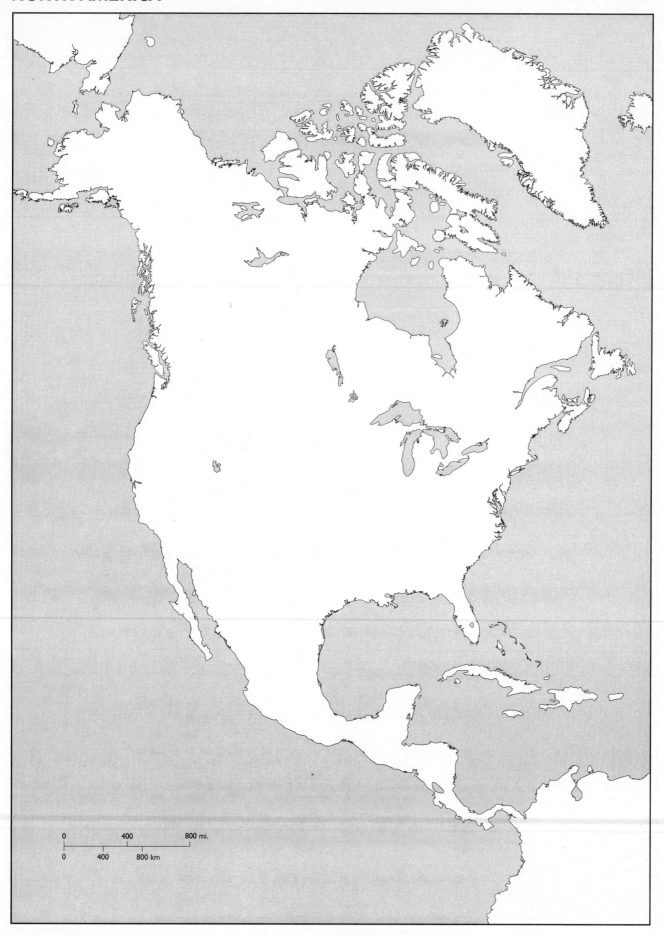

0   400   800 mi.

0   400   800 km

NORTH AMERICA

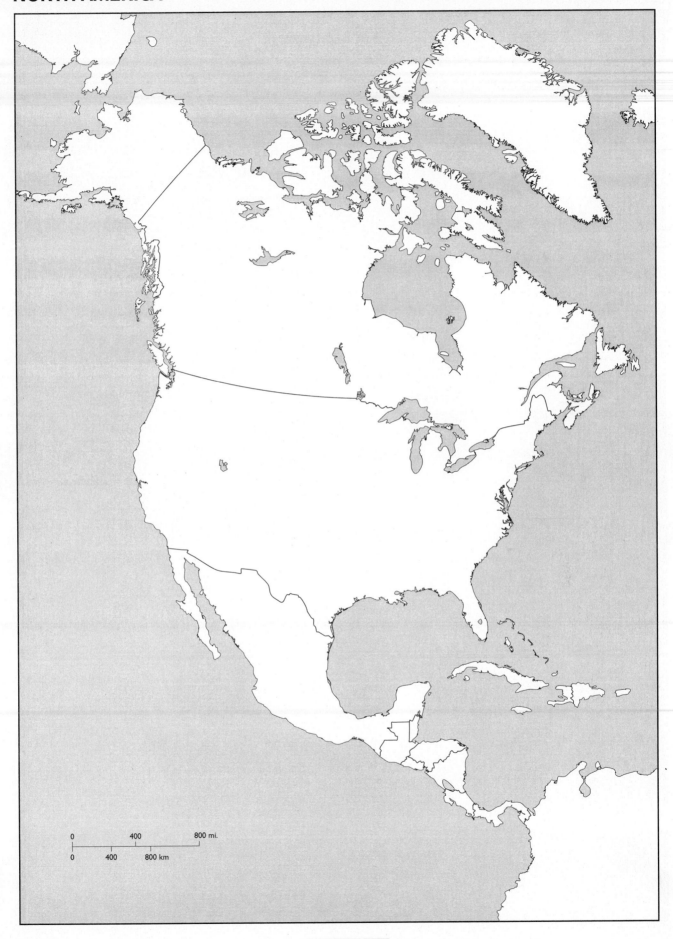

0 400 800 mi.

0 400 800 km

0   250   500 mi.

0   250   500 km

SOUTH AMERICA

0   250   500 mi.

0   250   500 km

SOUTH AMERICA

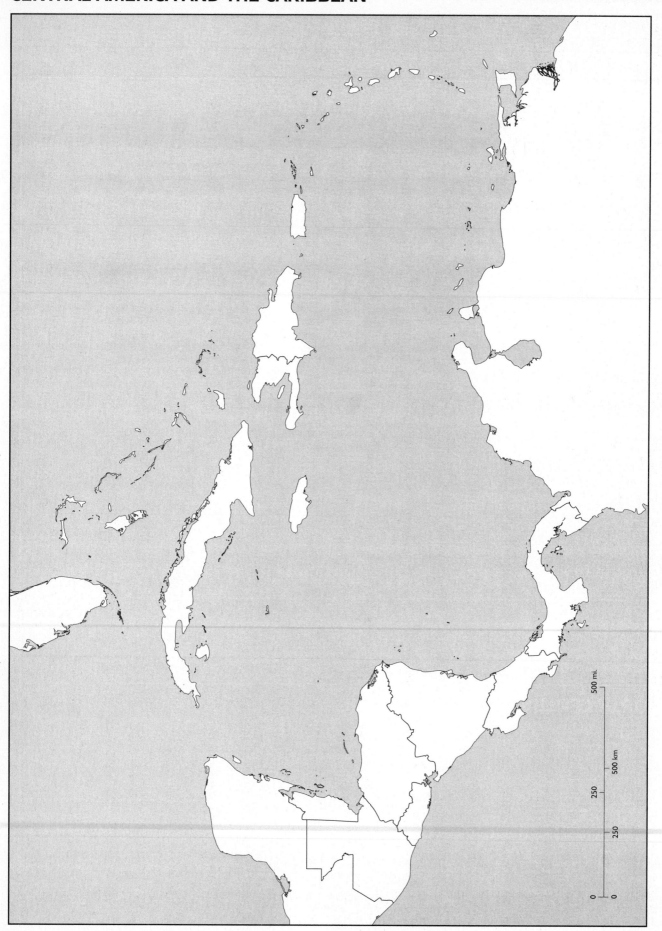

500 mi.

500 km

250

250

0

0

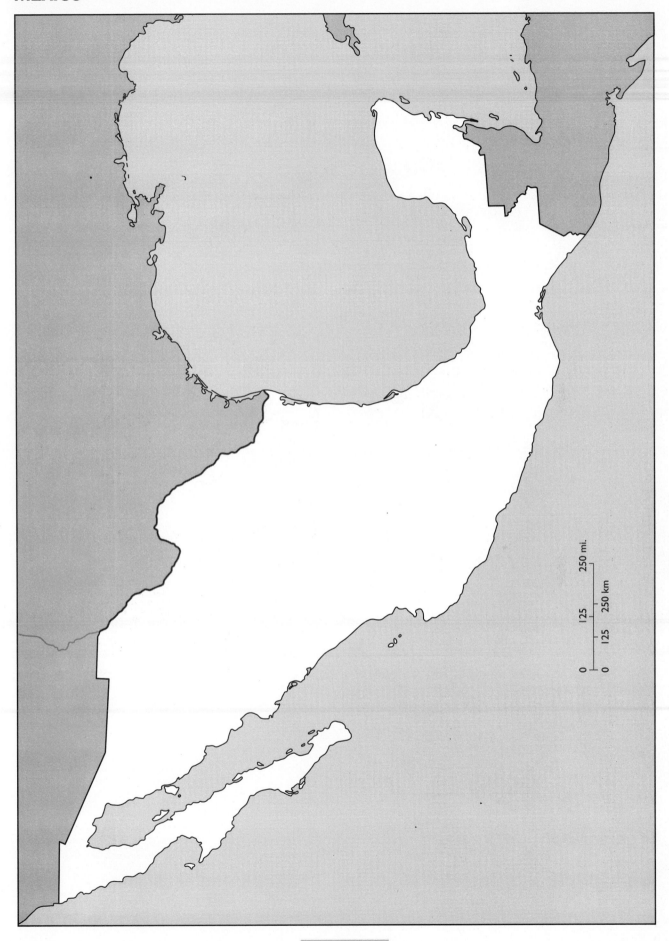

250 mi.

250 km

125

125

0

0

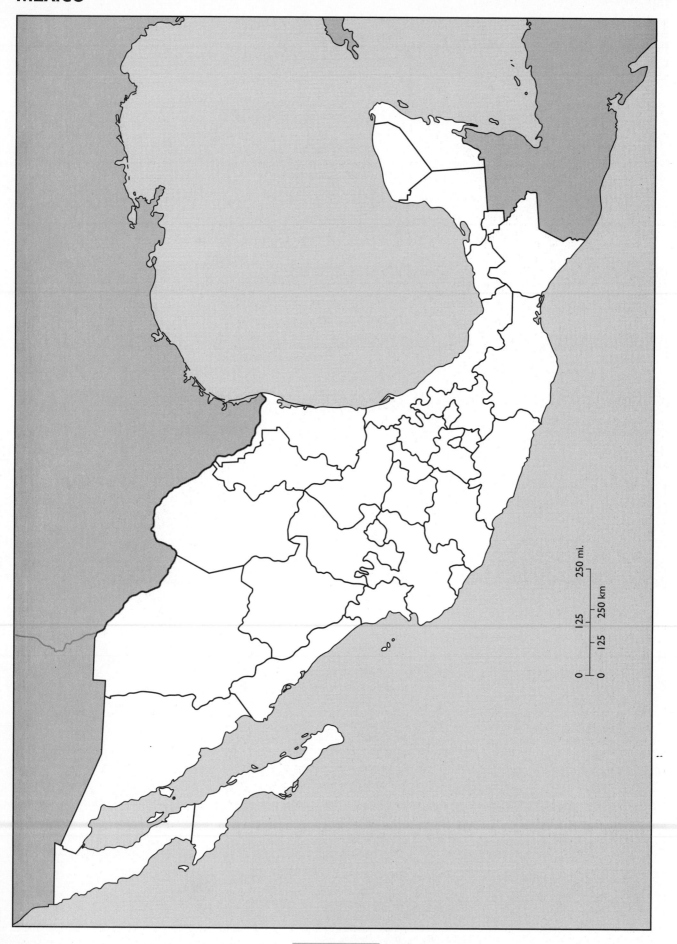

250 mi.

250 km

125

125

0   0

WESTERN HEMISPHERE

0   750   1,500 mi.

0   750   1,500 km

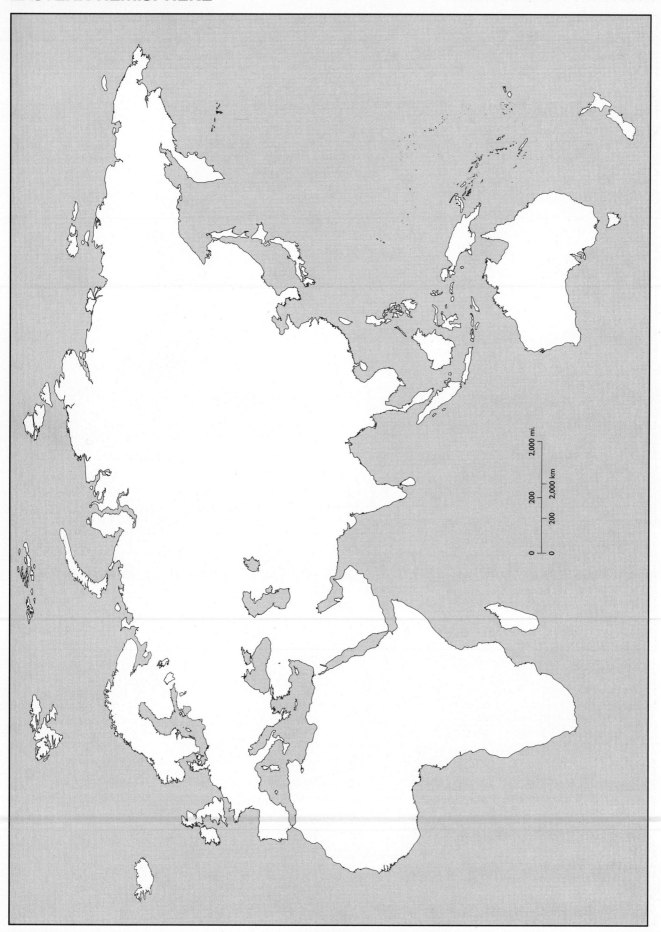

EASTERN HEMISPHERE

EUROPE

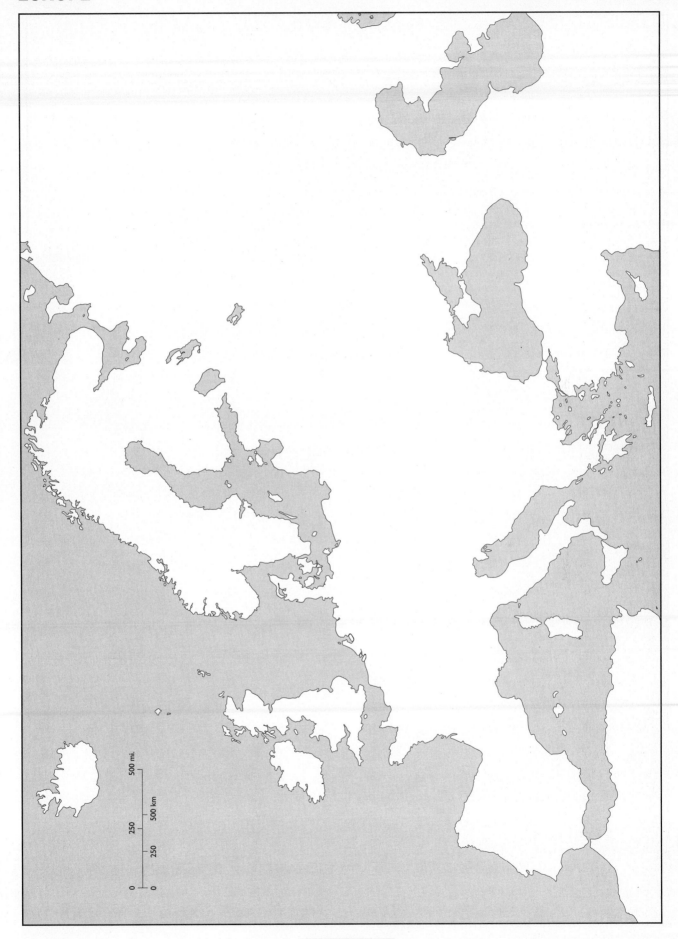

500 mi.

500 km

250

250

0

0

EUROPE

EUROPE

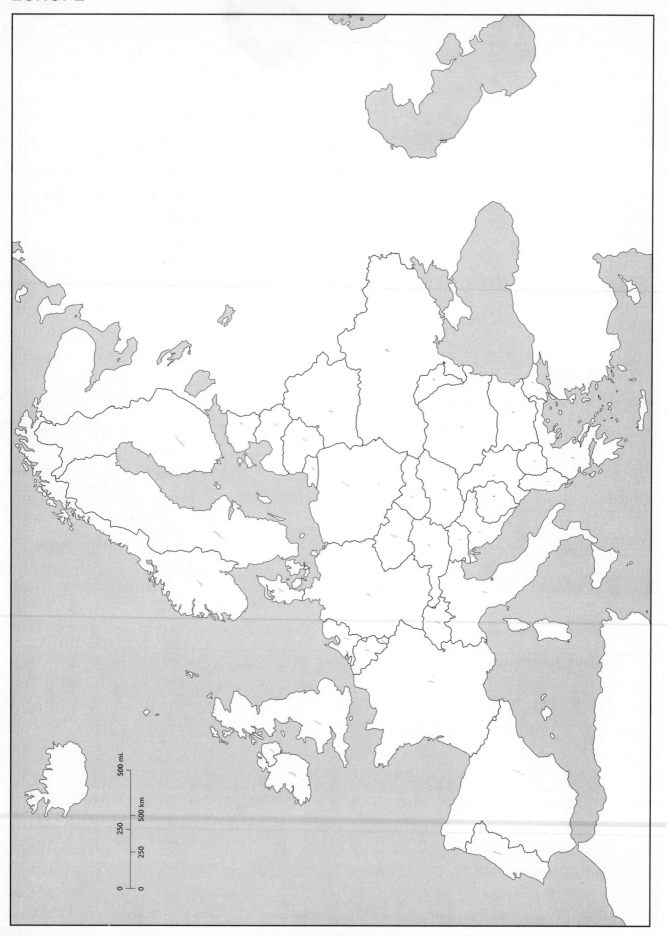

500 mi.

500 km

250

250

0 0

RUSSIA

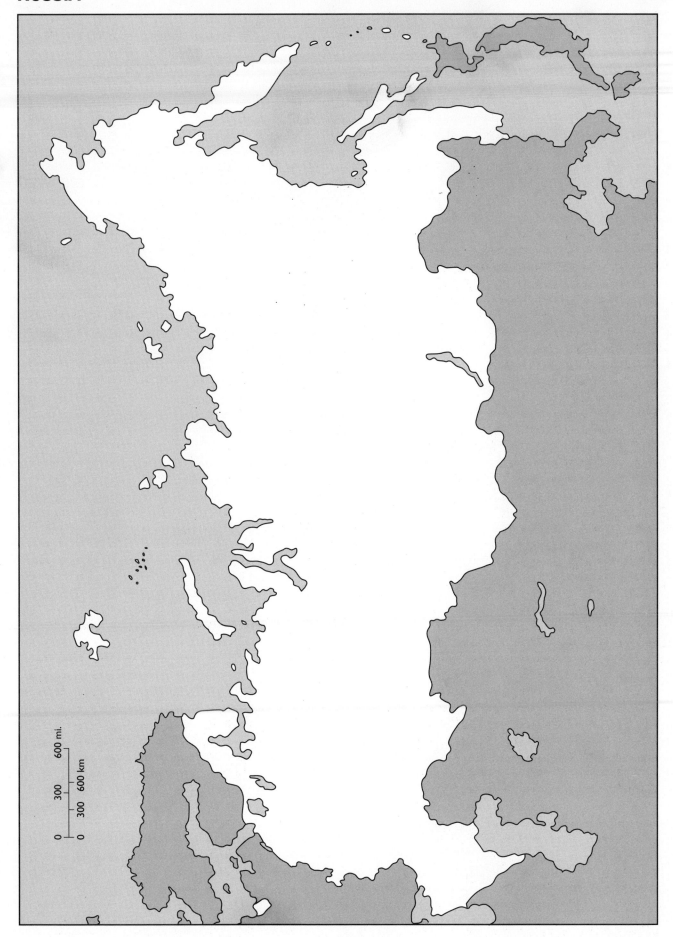

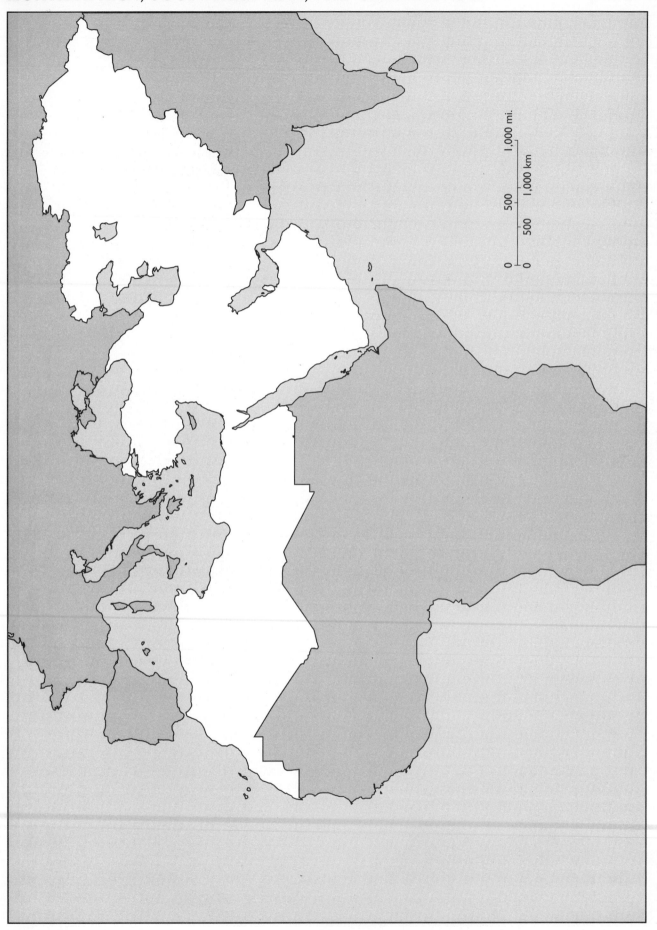

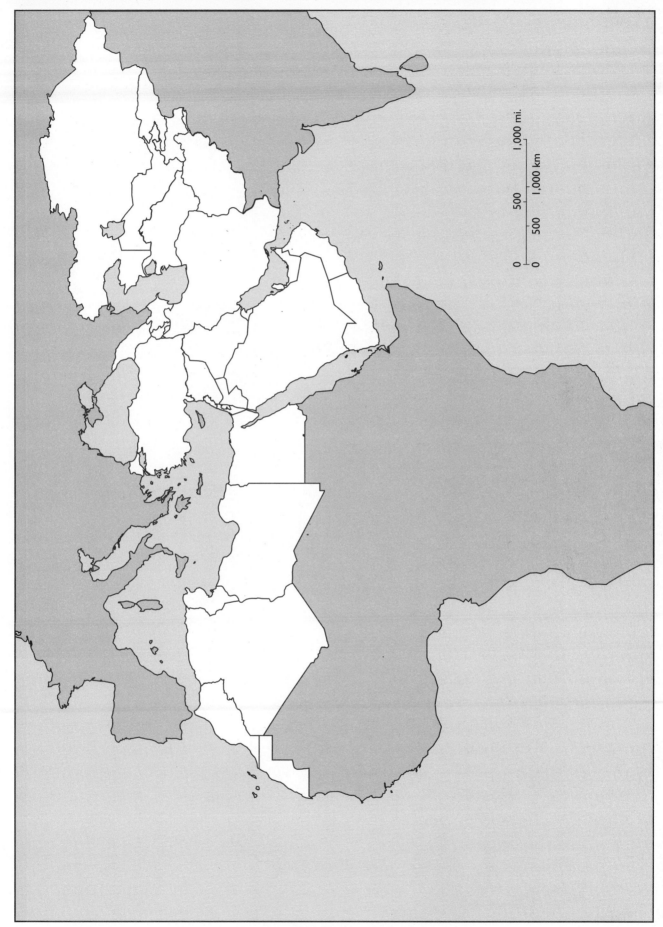

1,000 mi.

1,000 km

500

500

500

0

0

ASIA

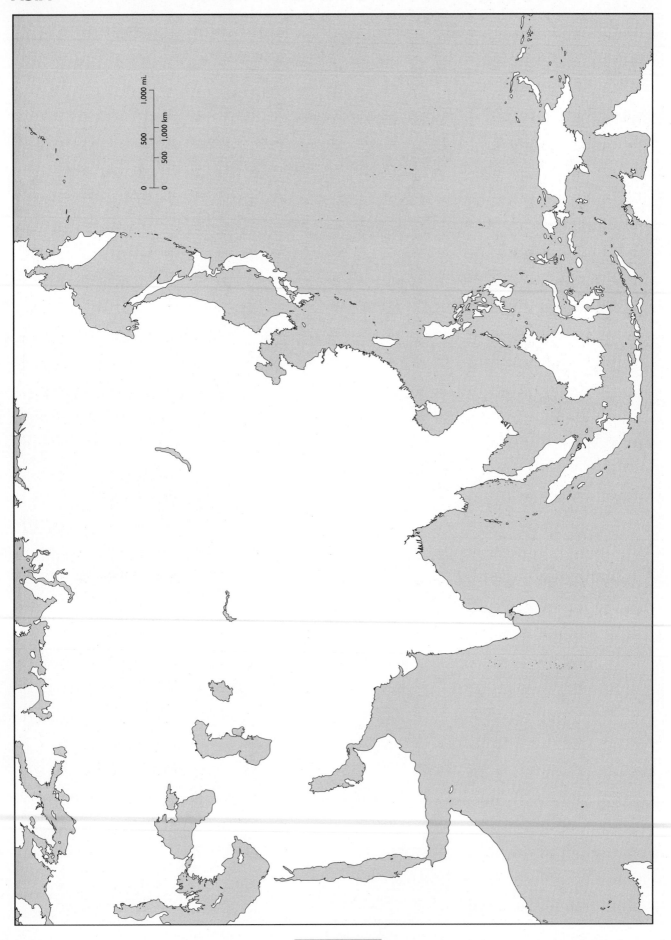

1,000 mi.

1,000 km

500

500

0

0

ASIA

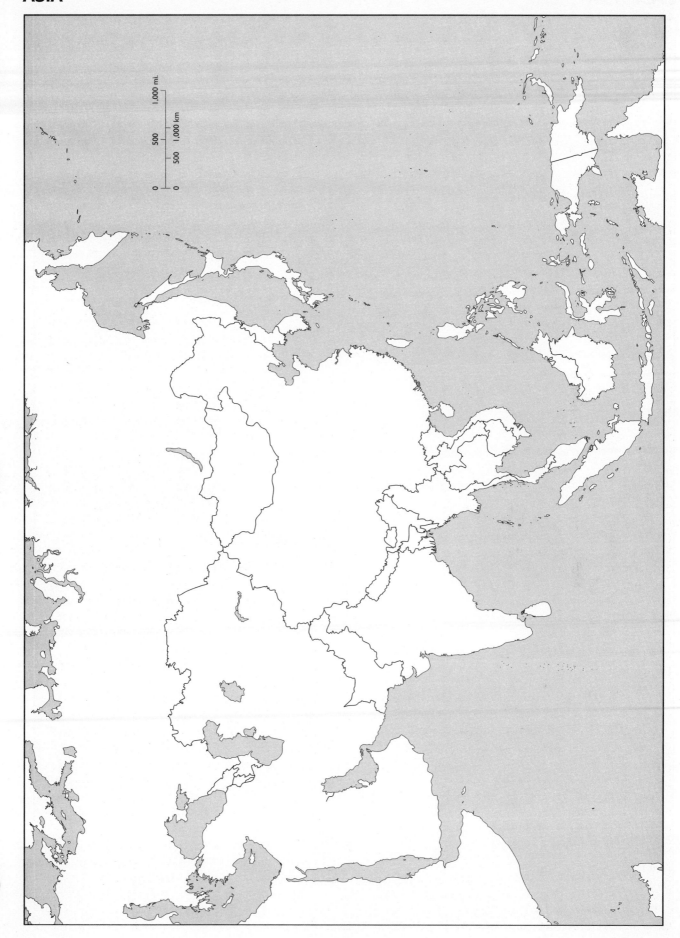

EAST ASIA

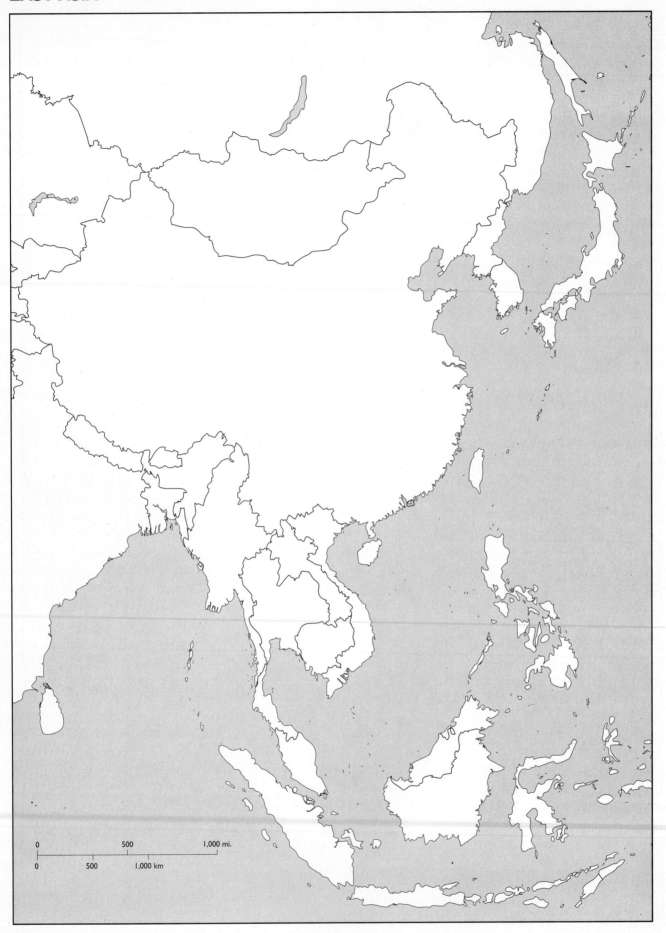

0 500 1,000 mi.

0 500 1,000 km

AFRICA SOUTH OF THE SAHARA

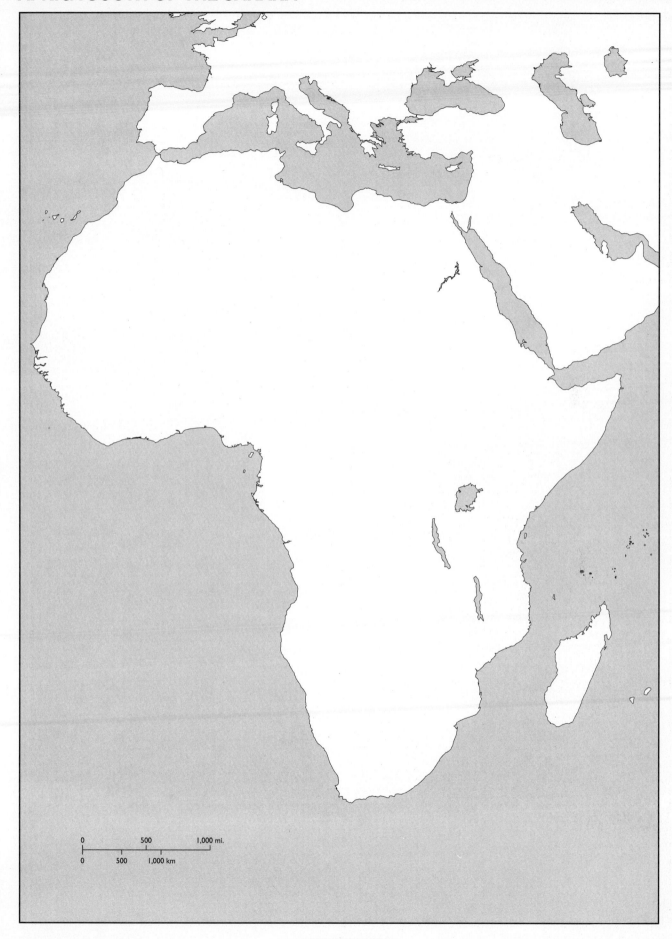

0 500 1,000 mi.

0 500 1,000 km

AFRICA SOUTH OF THE SAHARA

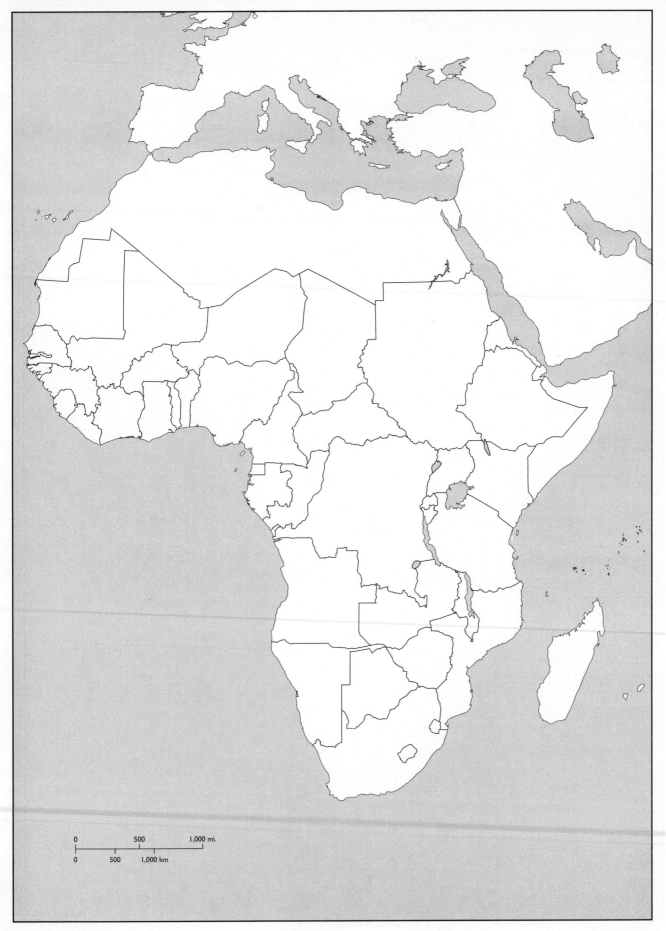

AFRICA SOUTH OF THE SAHARA

0   500   1,000 mi.

0   500   1,000 km

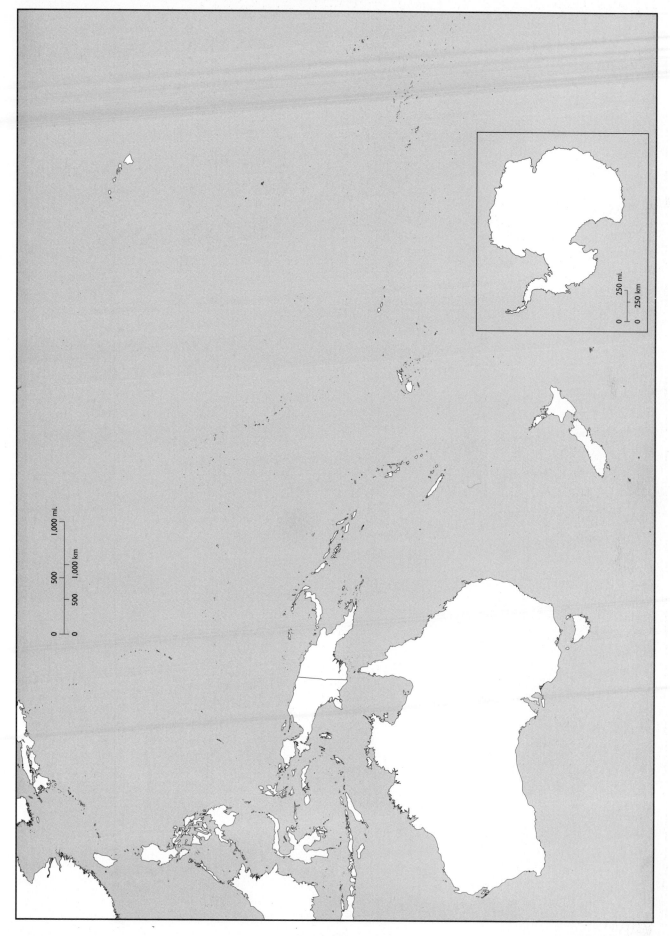

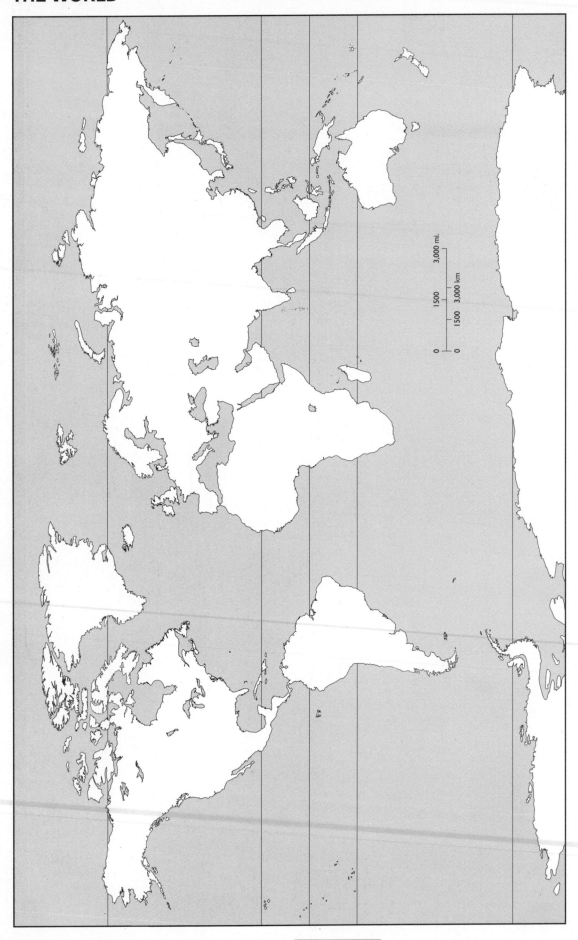

3,000 mi.

1500

0

3,000 km

1500

1500

0